AF688559

The Science of Neuroplasticity

Neuroplasticity & Recovery: How Your Brain Repairs Itself After Trauma or Stroke

By Joe Correa

Published by Live Stronger Faster

Copyright © 2025 by Joe Correa

All rights reserved.

No part of this publication may be reproduced, distributed, o transmitted in any form or by any means—electroni mechanical, photocopying, recording, or otherwise—withou the prior written permission of the publisher, except in th case of brief quotations embodied in critical reviews an certain other noncommercial uses permitted by copyright law

This book is for educational and informational purposes onl The author and publisher are not engaged in rendering lega financial, or other professional advice. Readers should consu a qualified professional before making any financial decisions

First Edition

Table of Contents

Introduction 5

Chapter 1 – Understanding the Basics of Neuroplasticity 8

Chapter 2 – The Science of Neurons and Synapses 13

Chapter 3 – Structural vs. Functional Plasticity 18

Chapter 4 – The Role of Repetition and Practice 23

Chapter 5 – The Emotional Brain and Neuroplasticity 28

Chapter 6 – Age and Neuroplasticity 33

Chapter 7 – Neuroplasticity in Learning and Skill Development 38

Chapter 8 – Neuroplasticity and Technology 43

Chapter 9 – Barriers to Neuroplasticity 48

Chapter 10 – Building a Brain for Life 53

Epilogue: The Living Brain 58

Introduction

The Science of Neuroplasticity

The human brain has often been described as the most complex structure in the universe. For decades, scientists believed that once brain cells were damaged or destroyed, recovery was impossible. People were told that strokes, traumatic brain injuries, or neurological disorders left behind permanent losses. Yet, modern neuroscience has revealed something extraordinary: the brain is not fixed but adaptable. This ability, known as neuroplasticity, allows the brain to reorganize itself, form new connections, and even assign functions to areas that were never originally responsible for them. What once seemed like permanent decline is now seen as a starting point for growth and repair. Neuroplasticity is not only a scientific concept—it is a message of hope.

Every year, millions of people worldwide suffer strokes, brain injuries, or trauma that alters their lives in profound ways. Families struggle to adjust, and survivors often face the daunting challenge of relearning the most basic skills—speaking, walking, remembering, or even smiling. The journey can feel lonely and overwhelming. This series was created to serve as a comprehensive guide for survivors, caregivers, and anyone interested in understanding the healing power of the brain. It gathers the latest science, prove recovery strategies, and real-world stories into one structure resource. Whether you are beginning recovery, supporting loved one, or simply curious about the brain's resilience, thi bundle gives you both knowledge and practical tools.

What Neuroplasticity Means for Recovery

To understand the significance of neuroplasticity, imagine road system where a bridge has collapsed. For a time, traffi may come to a halt, but gradually, detours are built, nev routes are discovered, and movement resumes. The brai works in much the same way. When one area is injured surrounding regions can take over lost functions. Neuron sprout new connections, and with repetition and effort, thes connections become stronger. This is why a stroke survivo may relearn how to walk or speak, even years after their injury Neuroplasticity proves that healing is not about going back t what was lost but about creating new pathways for what i possible.

A Journey of Hope and Renewal

Recovery after trauma or stroke is never easy. It demand patience, effort, and resilience. Yet, the fact that the brain ca adapt gives every survivor a reason to keep going. The patl may be slow, but every small improvement—a word spoken, hand lifted, a memory regained—is proof that healing is happening. This bundle is an invitation to believe in your brain's ability to recover and to take an active role in guiding that recovery. The goal is not simply survival but renewal: the chance to live fully again, perhaps with a deeper appreciation for the strength of the human spirit and the remarkable adaptability of the brain.

Chapter 1: Understanding the Basics of Neuroplasticity

Neuroplasticity is the brain's remarkable ability to chang adapt, and reorganize itself throughout life. For many year scientists believed the brain was fixed after childhood, wit little capacity for growth or recovery. That belief has bee overturned. Modern neuroscience has proven that the brai remains flexible, capable of building new pathways an modifying old ones in response to experience, learning, an injury. This discovery has transformed the way we approac trauma, stroke recovery, and lifelong learning.

The way neuroplasticity works can be compared to buildin and maintaining a network of roads. Neurons, the brain' communication cells, form connections through pathways th act like highways for information. When you practice a skill c repeat a thought pattern, it's as though more traffic flow along those roads, making them stronger and more efficien When connections are rarely used, they fade, like abandone paths that grow over with weeds. This principle explains wh repetition and practice are at the core of learning and recover and why change is possible at any age.

There are different forms of neuroplasticity, each playing role in adaptation and healing. Structural plasticity refers t the brain's ability to physically change its wiring by growing new synapses and reorganizing regions. Functional plasticity, on the other hand, describes how the brain shifts activity from one damaged region to another healthy one. Synaptic plasticity involves the strengthening or weakening of signals between neurons, making learning and memory possible. Compensatory plasticity allows the brain to develop alternative strategies to perform the same task when traditional pathways are damaged. Together, these forms create the foundation for recovery and growth.

The importance of repetition in shaping the brain cannot be overstated. Every time you repeat an action or practice a skill, you strengthen the pathways associated with it. This is why rehabilitation after a stroke or trauma involves consistent, repetitive tasks. A person relearning how to speak may repeat single words hundreds of times; a person working to regain mobility may practice small steps daily. Just as lifting weights builds muscles, repetition builds stronger neural circuits. Without it, recovery slows, but with persistence, improvements compound over time.

The environment surrounding a person greatly influences neuroplasticity. A stimulating, supportive, and enriched environment encourages the brain to adapt. Social interactions, problem-solving tasks, music, art, and new challenges all activate and strengthen neural connections. In contrast, environments that lack stimulation may limit progress. Caregivers and family members should recognize that their encouragement, patience, and active involvement are not merely emotional support—they are biological tools that shape the brain's recovery.

For individuals recovering from trauma or stroke neuroplasticity provides hope where once there was despair. It proves that healing does not end with survival. The brain is dynamic, always responding to input and effort. Rehabilitation exercises, therapy sessions, and daily practice are not just routines—they are opportunities to rewire the brain. Even years after injury, progress is still possible because the capacity for change never fully disappears. Understanding neuroplasticity reminds us that the brain is not static but a living, adaptable system, capable of creating new beginnings.

Daniel's wake up story

When Daniel first woke up after his stroke, the world felt unfamiliar. His right hand refused to move, and words that once flowed easily became tangled in his throat. He had been a confident lawyer, a man whose life revolved around precision and language, yet now even forming a sentence seemed impossible. He feared he would never return to the person he once was. The doctors explained that part of his brain had been damaged, but they also introduced a concept he had never heard before: neuroplasticity. His brain, they said, still had the power to change.

At first, the idea seemed far-fetched. Daniel could not imagine how a broken brain could repair itself. But his therapist assured him that the brain was not broken beyond repair—it was adaptable. Each attempt to move, speak, or remember, no matter how small, signaled the brain to build new connections. The therapist compared it to carving a trail through a forest: the more times you walk the path, the clearer it becomes. Daniel realized that his recovery would not be about miracles but about repetition and persistence.

The early days were full of frustration. He tried to say simple words, but they came out slurred. He tried to grip a cup, but it slipped from his hand. Yet his therapist encouraged him to keep going, reminding him that even the failed attempts were training his brain. Every effort was a signal. Slowly, progress appeared. One morning, he managed to button his shirt without help. Another day, he spoke an entire sentence without stumbling. These victories were small in the eyes of others, but to Daniel, they were milestones, proof that his brain was changing.

His family played an essential role in his progress. His wife filled their home with reminders of the life he loved: photos of their travels, music they once danced to, books that sparked conversation. His children encouraged him to practice by asking him to read aloud, even if the words came slowly. These moments filled his environment with stimulation, giving his brain countless opportunities to adapt. Daniel realized that recovery was not just about therapy sessions but about every interaction, every challenge, every attempt to connect.

Months passed, and Daniel's improvements grew more visible. He regained enough movement to write short notes, and his speech became clearer. He discovered that neuroplasticity was not a theory—it was happening inside him every day. His brain was reorganizing itself, rerouting functions to healthier regions, and strengthening new pathways with every repetition. He no longer measured progress by what he had lost but by what he was regaining.

Looking back, Daniel understood that his stroke had forced him to see the brain in a new light. It was not fragile but resilient, not fixed but adaptable. He realized that his identity was not lost; it was being reshaped alongside his brain. Recovery had become more than survival—it was a journey of renewal. When he spoke to other survivors, he shared his story, reminding them that neuroplasticity was not just science. It was hope, woven into the very fibers of their minds.

Chapter 2: The Science of Neurons and Synapses

The brain is built on a foundation of specialized cells called neurons, which are responsible for transmitting information across the nervous system. Each neuron is structured with a cell body that holds the nucleus, branching dendrites that receive incoming signals, and a long axon that sends signals outward. Together, they form intricate networks of communication that govern everything from movement and speech to memory and imagination. Without neurons, the brain's ability to adapt through neuroplasticity would not exist, because it is their connections that change and grow in response to experience.

Communication between neurons happens at junctions called synapses. A synapse is a tiny gap where an electrical signal traveling down the axon of one neuron triggers the release of chemicals called neurotransmitters. These molecules cross the gap and attach to receptors on the dendrites of the next neuron, sparking a new electrical impulse. This chain reaction repeats endlessly, forming the basis of thought and action. The strength of these synapses is not fixed; they can become more efficient with use or fade away without it. This ability to strengthen or weaken over time is known as synaptic plasticit and is at the heart of learning and recovery.

Different neurotransmitters carry out different functions, an each plays a role in neuroplasticity. Dopamine reinforce learning and motivation by rewarding the brain for successfu actions. Serotonin regulates mood and stabilizes emotions making recovery less overwhelming. Glutamate is the mos important neurotransmitter for learning and memory, whil GABA helps calm overactive neural signals. Acetylcholin sharpens attention and supports memory. When thes chemicals flow in balance, the brain is primed for healing an growth. When they are imbalanced, challenges such a depression, fatigue, or slowed recovery may occur.

One of the most powerful processes in the brain is long-terr potentiation, or LTP. This occurs when a synapse is repeatedl stimulated, making the communication between neuron faster and more reliable. Imagine practicing a word over an over during speech therapy. Each repetition strengthens th neural pathway, and eventually, recalling the word become effortless. LTP is considered the cellular basis of memory an learning. It demonstrates why consistency is so crucial i rehabilitation and why progress builds over time throug steady effort.

Neuroplasticity also depends on pruning, the process by which the brain eliminates connections that are no longer useful. Just as a gardener trims branches so that a tree can grow stronger, the brain refines its networks to function more efficiently. This means that repetition not only strengthens important pathways but also allows the brain to discard weak or unnecessary ones. In recovery, this process can be a double-edged sword: consistent practice leads to stronger recovery, while neglect can cause skills to regress.

Understanding neurons and synapses provides more than scientific knowledge—it gives survivors of trauma or stroke practical insight into why recovery takes effort and persistence. Every therapy exercise, every repeated attempt at movement, and every word practiced is not wasted. Each one signals neurons to fire, release neurotransmitters, and reinforce synapses. This process proves that healing is not about chance but about biology. By engaging in daily practice, patients take control of their recovery, guiding their neurons to build the connections they need for a fuller life.

Emily's concussion

When Emily woke up in the hospital after a serious concussion, she felt like her mind was foggy and disconnected. She struggled to focus on conversations and often forgot what someone had said just minutes earlier. The doctors told her that her neurons had been jolted by the injury, but they also explained that her brain was not permanently broken. They told her that neurons and synapses could adapt and rebuild and that her job would be to give them the right signals through practice and patience.

In her early therapy sessions, Emily worked on simple memory exercises. She would look at a series of pictures and then try to recall them. At first, it was discouraging; she could barely remember two or three images. Her therapist reminded her that every attempt was like exercising a muscle. The neurons needed repetition to strengthen their connections. Though she doubted herself, Emily decided to keep trying, clinging to the idea that her brain was capable of more than she could see in that moment.

Over time, she began to notice small improvements. She could remember longer lists and hold conversations without losing her train of thought. What she didn't realize was that long-term potentiation was at work. Every time she practiced, her neurons were firing in the same patterns, releasing neurotransmitters that made those pathways more efficient. Her brain was literally rewiring itself, and the fog slowly began to lift.

Her therapist introduced emotional cues into her exercises, such as asking her to recall words that were tied to happy memories. When Emily saw a photo of her childhood dog, she not only remembered the word "dog," but also the smell of the beach where she used to play fetch. These emotional triggers released dopamine and serotonin, strengthening her synapses even more. She discovered that recovery was not only about practice but also about finding meaning in what she practiced.

Months later, Emily reached a milestone. She sat with her best friend and recited the details of a story they had both experienced years before. Her friend's eyes filled with tears as she listened, knowing how far Emily had come since those early days when she could barely recall a single sentence. Emily realized then that every frustrating exercise, every repeated attempt, had been carving out new roads in her brain. The effort had been worth it.

Looking back on her journey, Emily understood that neurons and synapses were not just scientific terms—they were the foundation of her recovery. She had experienced firsthand the way they adapt, grow, and strengthen when given the right signals. Her story became a source of encouragement for others facing similar struggles. She often told them, "Your brain is always listening. Every attempt you make is a message

to your neurons to keep building. Don't stop sending thos
signals."

Chapter 3: Structural vs. Functional Plasticity

Neuroplasticity expresses itself in more than one way. The two most widely studied forms are structural plasticity and functional plasticity. Structural plasticity refers to the brain's ability to physically change its wiring. This includes the growth of new dendrites, the formation of new synapses, and even the reorganization of gray matter in response to experience or injury. Functional plasticity, by contrast, is the brain's ability to shift tasks from damaged areas to healthier regions. Both processes work together to make recovery after trauma or stroke possible.

Structural plasticity can be thought of as construction at the cellular level. When you practice a new skill or repeat a therapeutic exercise, the brain begins to build new connections to support that activity. Over time, these new pathways become stronger and more efficient, allowing the skill to be performed with less effort. Brain scans of people who learn to play an instrument, for example, show measurable increases in gray matter density in regions associated with motor control and auditory processing. This physical remodeling demonstrates that experience leaves a visible imprint on the brain.

Functional plasticity is different. Instead of building ne structures, it reassigns duties when certain regions can n longer perform them. If an injury damages the left hemisphe where language is typically processed, the right hemisphe may take on some of that function. This reassignment may n restore abilities exactly as they were before, but it provides pathway for regaining independence. Functional plasticity especially important in the early stages of recovery, when th brain is searching for alternative routes to bypass damage areas.

Modern imaging techniques such as fMRI have given us clearer picture of these processes. Researchers can now se how brain activity shifts after injury, often to regions that we not previously responsible for certain tasks. In strok survivors, for example, increased activity in the undamage hemisphere has been observed as they relearn how to speak c move. These scans offer proof that the brain is not fixed bu dynamic, always searching for solutions.

Rehabilitation strategies often target both structural an functional plasticity. Repetition of movements or word encourages the brain to physically strengthen pathways, whi problem-solving and adaptive exercises challenge the brain t reassign tasks to other regions. This is why therapies a designed to be both repetitive and varied. They provide th stimulation necessary to build new structures while encouraging the brain to experiment with different solutions.

Understanding the difference between structural and functional plasticity is critical for setting realistic expectations during recovery. Structural changes take time because they involve physical growth and remodeling, while functional changes can occur more quickly but may not restore abilities to their original state. Both, however, are essential. Together they explain why progress is often uneven, with sudden breakthroughs followed by slower periods. Recognizing this balance allows patients and caregivers to appreciate that every effort contributes to the brain's remarkable ability to adapt.

Stroke at 62

When Samuel suffered a stroke at sixty-two, he feared his painting career was over. His right hand trembled uncontrollably, and the brush that had once felt like an extension of his body now slipped helplessly from his grasp. His doctors explained that the stroke had damaged regions of his brain that controlled fine motor skills. But his therapist reminded him that his brain still had the capacity to adapt, not only by growing new pathways but also by shifting functions to healthier areas.

At first, Samuel tried to practice with his injured hand, moving it clumsily over paper. The lines were shaky, and frustration threatened to consume him. Yet every small effort was stimulating structural plasticity. Even though he couldn't see it, his brain was beginning to form new connections, sprouting tiny networks that would eventually support better control His therapist encouraged him to keep repeating the movements, reminding him that each attempt was laying the groundwork for recovery.

Meanwhile, Samuel discovered something unexpected. When he picked up a brush with his left hand, he was able to create simple sketches. They were not as precise as his earlier work but they carried a unique style. His therapist explained tha this was functional plasticity in action. His brain was reallocating artistic control to his non-dominant hand allowing him to continue expressing himself while his righ hand healed. It wasn't a replacement, but it was a bridge tha gave him hope.

Weeks turned into months, and Samuel's scans showed increased activity in areas of his brain that had not been involved in painting before the stroke. His family wa astonished when he produced his first full painting, a vibran abstract piece created with both hands. The work was differen from his earlier style—bolder, less detailed—but it carried a depth that reflected his journey. His brain was learning to combine both structural growth and functional reassignment to reclaim his artistry.

The breakthrough reinvigorated Samuel's determination. Every day, he practiced for hours, repeating basic strokes with his right hand while experimenting with new techniques using his left. Slowly, his control improved. His therapist explained that structural plasticity was strengthening the connections in his right hand while functional plasticity continued to provide support from other regions. The combination was giving him more independence than he had ever expected.

Years later, Samuel began teaching art classes for other stroke survivors. He used his story as proof that recovery was not about going back to who you were but about discovering new possibilities. He told his students that the brain was both a builder and an adapter, capable of growing new structures and shifting old ones. "Your art may look different," he said, "but that doesn't mean it's less. It means your brain is still creating with you." His story became a living example of structural and functional plasticity at work, inspiring others to believe in their own recovery.

Chapter 4: The Role of Repetition and Practice

Repetition is one of the most powerful forces drivin neuroplasticity. Each time you repeat an action, whether it i lifting a hand, speaking a word, or practicing a movement, yo are sending signals across specific neural pathways. Thes signals act like footsteps on a trail. The more you walk th trail, the clearer it becomes, until it eventually forms into well-paved road. In the brain, this means the pathwa becomes faster, stronger, and more efficient. Withou repetition, neural connections remain weak, but with practice they gain strength and stability.

The principle behind repetition is often summarized a "neurons that fire together, wire together." When neurons ar activated at the same time, their connections become mor synchronized. Over time, these repeated activations make i easier for the brain to recall an action or memory. Thi explains why skills that are practiced frequently becom automatic, such as tying shoelaces or riding a bicycle. Fo individuals in recovery, it also explains why consistent practic is essential for regaining lost abilities.

Repetition must be deliberate to have the greatest effec Passive activity, such as listening without engagement, doe not build connections in the same way. Deliberate practice requires focused attention, effort, and feedback. A person practicing speech recovery, for example, benefits more from repeating words while concentrating on pronunciation than from casual conversation alone. In this way, quality matters as much as quantity when it comes to rewiring the brain.

The effects of repetition extend beyond movement and speech. Memory, decision-making, and emotional regulation are also strengthened through repeated practice. Exercises such as recalling daily events, practicing gratitude, or engaging in mindfulness all reinforce circuits that support mental health and resilience. This demonstrates that repetition is not limited to physical recovery but is equally important for cognitive and emotional healing.

Repetition also plays a role in pruning. When useful pathways are repeated, they grow stronger, while pathways that are ignored gradually weaken and fade away. This dual process helps the brain become more efficient, focusing energy on skills that are reinforced while letting go of those that are no longer relevant. In rehabilitation, this means that consistent effort not only strengthens recovery but also prevents regression by keeping important pathways active.

Perhaps the most inspiring aspect of repetition is that progress compounds over time. Early attempts may feel frustrating and unproductive, but each effort contributes to building the foundation for future success. Improvements may seem invisible at first, but they accumulate until breakthroughs occur. This truth reminds patients and caregivers that persistence matters more than perfection. Every small repetition is an investment in the brain's ability to change, creating possibilities that extend far beyond what seems possible in the moment.

Olivia's speech therapy

When Olivia first began speech therapy after her stroke, she could only manage a few sounds. Words seemed trapped in her mind, unreachable no matter how hard she tried. Her therapist handed her a card with the word "sun" on it and encouraged her to repeat it. The first attempt came out garbled. She tried again, her voice strained and uneven. By the tenth attempt, she wanted to give up, but her therapist reminded her that each try was helping her brain relearn the path from thought to speech.

The early days were discouraging. Progress was so slow that Olivia often doubted whether she would ever recover her voice. But she kept returning to the same exercises: repeating words, forming sentences, and practicing sounds in front of a mirror. Day after day, she walked the same mental trail, not realizing that her neurons were wiring themselves more tightly together with each repetition.

One afternoon, several weeks into therapy, Olivia surprised herself. She read through a list of words and realized that they came out more clearly than before. The improvement was not dramatic, but it was undeniable. Her therapist explained that repetition had strengthened the connections between the neurons involved in speech. Olivia had been planting seeds every day, and now they were beginning to grow.

Encouraged, she began practicing outside of therapy. She labeled objects in her kitchen and spoke their names aloud whenever she saw them. She read short stories to her granddaughter, struggling through the sentences but refusing to stop. Each attempt was like a workout for her brain, and slowly, the fog began to lift. Her family noticed the difference, celebrating each milestone with her, no matter how small.

Months later, Olivia delivered a toast at her daughter's birthday party. Her voice was not as smooth as it had once been, but she spoke in full sentences, clear enough for everyone to understand. Tears filled her eyes as she raised her glass. The repetition that once felt like endless struggle had built the foundation for this moment. She realized that every attempt, every word practiced in isolation, had been paving the way toward recovery.

Looking back, Olivia understood that repetition had given he more than her speech—it had given her patience, discipline and hope. Her journey taught her that progress was not abou sudden leaps but about steady steps taken over and over again She began mentoring others in recovery, telling them that eve when practice feels tedious, it is shaping the brain in ways tha cannot always be seen in the moment. “Keep trying,” she said “Every effort counts, even the ones that feel like failures. You brain is always listening.”

Chapter 5: The Emotional Brain and Neuroplasticity

Emotions play a far greater role in neuroplasticity than most people realize. While repetition and practice form the structure of learning and recovery, emotions provide the energy that powers those processes. The brain is not a machine that works in isolation; it responds intensely to feelings of joy, fear, frustration, motivation, and love. These emotions activate neurotransmitters and hormones that influence how strongly neural pathways form or fade. Positive emotions can accelerate learning and recovery, while chronic stress and negativity can slow or even block progress.

The limbic system, which includes the amygdala, hippocampus, and parts of the prefrontal cortex, governs emotional processing. The amygdala is responsible for detecting threats and generating fear responses, while the hippocampus helps convert experiences into memories. When we feel emotionally connected to something—whether through excitement, curiosity, or empathy—the hippocampus becomes more active, encoding those experiences more deeply. This is why emotionally charged memories tend to stay with us longer, and why emotional engagement is so powerful during therapy.

Stress, however, can disrupt neuroplasticity. When the brai perceives danger, it releases cortisol, a stress hormone tha prepares the body for survival. Short bursts of stress ca sharpen focus, but chronic stress has the opposite effect. Hig cortisol levels reduce the brain's ability to form ne connections and can even damage neurons in th hippocampus. For those in recovery, this means that fea frustration, and anxiety can interfere with healing. Learning t manage emotional stress is therefore as essential t neuroplasticity as physical or cognitive exercise.

Positive emotions have the opposite effect, promoting growt and repair. Dopamine, serotonin, and oxytocin—often calle "feel-good" neurotransmitters—enhance motivatio optimism, and connection. When therapy sessions are frame with encouragement, humor, or personal meaning, the brai becomes more receptive to change. For example, a strok survivor who practices speech exercises while talking with loved one experiences both repetition and emotiona stimulation, which together amplify results. Emotion engagement gives the brain a reason to care about the activit making it more likely to remember and improve.

This emotional influence is also why relationships and soci support are so important during recovery. Human connectio provides a steady stream of positive reinforcement that help sustain motivation when progress feels slow. Isolation, by contrast, can dull the brain's responsiveness and lead to feelings of hopelessness that undermine recovery. The presence of empathy, encouragement, and genuine care can often reignite neural growth even after long plateaus. Healing, therefore, is not just a physical or cognitive process but a profoundly emotional one.

Understanding the relationship between emotion and neuroplasticity changes how we approach recovery. It reminds us that healing requires more than discipline—it requires compassion, joy, and patience. Emotional states shape biology by influencing how the brain rewires itself. Every smile, every moment of gratitude, and every small success releases chemicals that strengthen new connections. By nurturing positive emotions and managing stress, we create the ideal environment for the brain to grow, adapt, and thrive.

Rehab for Marcus

When Marcus began his rehabilitation after a motorcycle accident, he was consumed by frustration. Once a confident and athletic man, he now struggled to lift a spoon or tie his shoes. Every session with his therapist felt like a reminder of what he had lost. His anger was constant, and his progress seemed painfully slow. One day, his therapist stopped the exercise and said softly, "Marcus, your body follows your brain, and your brain follows your emotions. Let's sta working on those first."

At first, he didn't understand what she meant. He thoug therapy was about movement, not feelings. But over time, h began to see how his anger and despair drained h motivation. The therapist introduced mindfulness techniqu to help him calm his thoughts before each session. H practiced deep breathing, focusing on gratitude for sma victories—like the fact that he could move his fingers agai Gradually, he noticed that when he approached therapy with calmer mind, his body responded better. The exercises th once felt impossible started to improve little by little.

One day, during a particularly difficult session, his wi entered the room carrying their newborn daughter. Marc had been afraid to hold her since the accident, unsure if he ha the control to do it safely. The therapist guided him throug the motions, and with trembling hands, he lifted the baby in his arms. The moment filled him with emotion—fear, joy, ar love all at once. His wife smiled, tears in her eyes. Th experience became a turning point. The emotional intensity the moment seemed to unlock something in his brain, makir the movements easier in the weeks that followed.

Marcus began to understand that emotions were not obstacles—they were tools. His therapist explained that when he felt connected, hopeful, and engaged, his brain released dopamine and serotonin, which improved his focus and motor control. Encouraged by this, Marcus started each session by listening to his favorite music, using the rhythm to guide his movements. The songs that once played in the background of his workouts now became the soundtrack to his recovery.

Months passed, and Marcus's improvement accelerated. He laughed more, connected more deeply with his family, and began mentoring others in therapy. He noticed that when he encouraged fellow patients, his own motivation grew stronger. The act of helping others gave him purpose, which in turn fueled his brain's plasticity. Emotion had transformed from a source of pain to a source of power.

Looking back, Marcus realized that recovery had been as much emotional as it was physical. His brain had learned to heal because he had learned to hope. The anger that once held him back had been replaced by gratitude and joy, and those emotions had reshaped his brain from the inside out. When he spoke at a rehabilitation event years later, he told the crowd, "The brain heals best when the heart leads the way." His journey became a living example of how emotion fuels neuroplasticity, turning struggle into strength.

Chapter 6: Age and Neuroplasticity

For many years, scientists believed that the brain's ability to change and adapt declined sharply after childhood. The common belief was that once the brain matured, it became fixed, with little room for new growth. Modern neuroscience has shattered that misconception. Today, we know that neuroplasticity continues throughout life. While it may occur more rapidly in youth, adults and even older individuals retain the ability to rewire their brains through learning, practice, and recovery. This discovery has redefined how we view aging, healing, and lifelong learning.

In children, neuroplasticity is at its peak. The young brain is like soft clay—malleable and ready to adapt. Every experience, from learning language to developing social skills, sculpts neural networks that define future behavior and ability. Children recover from brain injuries more quickly because their neural circuits are still forming and can easily reorganize around damaged areas. However, this flexibility comes with less stability, which is why habits—both good and bad—form so easily during childhood.

Adults experience a different kind of plasticity. While the brain becomes more stable with age, it retains the capacity for targeted growth. Adults may need more repetition and focused effort to create lasting change, but they also benefit from th structure and knowledge that experience provides. The adul brain can learn new languages, master new skills, and recove from injuries by stimulating specific pathways throug deliberate practice. The key is consisten engagement—keeping the brain challenged with meaningfu and varied activities.

In older adults, neuroplasticity does not disappear, but i slows. This slowdown is influenced by factors such as reduce blood flow, inflammation, and less frequent ment stimulation. Yet studies show that seniors who stay mentall and socially active maintain sharper cognition and bette emotional health. Activities such as reading, solving puzzle learning instruments, or exercising the body can all enhanc brain connectivity. In fact, physical movement is one of th most effective ways to boost neuroplasticity at any age, as i increases oxygen, circulation, and the release c growth-promoting hormones in the brain.

Emotional health also plays a critical role in maintainin neuroplasticity as we age. Depression, loneliness, and chroni stress can weaken neural connections and reduce the brain' ability to form new ones. Conversely, social connectio purpose, and curiosity stimulate the brain to keep growin The more an individual engages in meaningful relationship and activities, the more their brain continues to adapt. It is not age that stops neuroplasticity—it is inactivity.

Understanding that neuroplasticity is lifelong empowers people at every stage to invest in their mental and physical growth. Whether through learning new skills, staying socially connected, or practicing gratitude, the brain remains capable of transformation. The message is clear: while we cannot stop time, we can influence how the brain ages. Every new challenge, every moment of curiosity, and every act of persistence strengthens the networks that define who we are. The aging brain is not in decline—it is still a work in progress.

Cognitive decline, or not?

When Eleanor turned eighty, she began to forget small things—where she placed her keys, what day it was, or whether she had taken her morning medicine. Her doctor mentioned the possibility of mild cognitive decline, a phrase that frightened her. But instead of surrendering to the idea that her best days were behind her, she decided to test what her brain could still do. She enrolled in an art class at her local community center, something she had always wanted to try but never had time for.

At first, the lessons were difficult. Her hands trembled slightly as she tried to hold a paintbrush steady, and she often mixed up colors. But she kept practicing. Each class challenged he brain to coordinate movement, memory, and creativity. He instructor encouraged her to focus not on perfection but o expression. Over time, her brushstrokes became mo confident, and her paintings began to take shape. She notice that she could remember color combinations and technique more easily each week. Her brain, she realized, was learnin again.

Outside of class, Eleanor made other changes. She joined book club, began taking daily walks, and practiced ligh exercises to improve her balance. The more she engaged wit others, the more alert she felt. Her daughter remarked that he conversations had become livelier and that she seemed mo energetic. Eleanor didn't know the scientific explanation, bu she could feel something inside her shifting. Her world wa expanding, not shrinking.

During one of her doctor's visits, her physician was surprise to see that her memory tests had improved. He explained tha the activities she was doing—painting, reading, exercising, an socializing—were all stimulating neuroplasticity. New neura pathways were being formed, and old ones were bein strengthened. Eleanor smiled when she heard this. She ha always believed that curiosity kept the mind young, and no she had proof.

Months later, Eleanor organized a small art exhibit with her classmates. Friends, family, and neighbors came to see their work. As she stood beside her paintings, she reflected on how far she had come since that first shaky brushstroke. Learning to paint had not only improved her coordination and focus but had reignited her confidence and sense of joy. Her mind felt alive again.

In her later years, Eleanor became an advocate for lifelong learning at the senior center. She often told newcomers, “Your brain doesn’t stop growing just because you’ve gotten older. It just needs you to give it something to grow toward.” Her story became a beacon of hope for others who feared aging meant decline. Through her determination, she proved that age does not end neuroplasticity—it simply invites a new chapter of discovery.

Chapter 7: Neuroplasticity in Learning and Skill Development

Learning is one of the most powerful demonstrations of neuroplasticity. Each time we acquire a new skill—whether it's playing an instrument, mastering a language, or solving a complex problem—the brain physically changes. Neurons form new connections, existing pathways strengthen, and regions of the brain reorganize to support the new activity. This process doesn't just make us more skilled; it literally reshapes our brain's structure. Every act of learning is an act of self-rewiring, proving that growth is not only psychological but biological.

When we begin learning something new, the brain is inefficient. Neural signals must travel through unfamiliar routes, and errors are common. But as we practice, certain pathways are reinforced while others are pruned away. The brain starts to optimize performance by strengthening the most useful circuits and eliminating redundant ones. This is why repetition, consistency, and feedback are essential to learning. The more often a skill is practiced, the faster and more automatic it becomes. With time, the brain reorganizes itself to perform that task with minimal conscious effort.

Different kinds of learning engage different parts of the brain, but all rely on the same principles of neuroplasticity. Motor learning, such as dancing or writing, strengthens the cerebellum and motor cortex. Cognitive learning, such as studying mathematics or language, enhances the prefrontal cortex and temporal lobes. Emotional learning, like empathy and resilience, shapes the limbic system. These processes work together, making learning a whole-brain experience. What we repeatedly do becomes embedded not only in memory but in the brain's very architecture.

Feedback is another critical factor in neuroplasticity. When we receive feedback—whether from a teacher, a coach, or a device—it helps the brain evaluate performance and adjust its neural pathways accordingly. Corrective feedback strengthens accurate connections and weakens errors. This is why trial and error is so effective in skill acquisition: mistakes trigger the brain to search for better solutions. With each iteration, neural circuits refine themselves, gradually leading to mastery.

Motivation also plays an essential role in learning. The neurotransmitter dopamine drives curiosity and rewards effort, creating a sense of satisfaction that encourages continued practice. When learning is tied to purpose or passion, dopamine levels rise, accelerating the formation of new connections. This means that the most effective learners are not those who work the hardest but those who fin meaning in their efforts. The brain thrives when learning feel relevant, enjoyable, and connected to a larger goal.

Understanding the connection between learning an neuroplasticity allows anyone to improve more efficiently. I removes the illusion of fixed intelligence or talent and replace it with the truth of adaptability. The brain is a dynamic orga that rewires itself in response to engagement, focus, an repetition. Whether you are learning to play the piano, speak new language, or recover from a neurological injury, th principle is the same: every repetition, every challenge, an every moment of focused practice changes the brain at its core

The power of a violin

When Adrian decided to learn the violin at forty-five, most c his friends thought it was a passing hobby. He had no musica background, no training, and little free time between work an family. But after years of listening to his daughter practice, h became fascinated by how effortlessly she played. H wondered if his own brain, well into adulthood, could sti learn something completely new. His first lesson prove humbling. The bow screeched across the strings, his timin was off, and his fingers refused to cooperate. Yet his teache reassured him: “Your brain just doesn’t know the path ye Give it time—it will learn.”

Adrian began practicing thirty minutes a day. At first, it felt hopeless. His fingers ached, and his coordination lagged behind his intentions. But as the weeks went by, something started to change. The sounds became less chaotic, his timing steadier. He realized that his brain was slowly adjusting. Each practice session felt like a small battle between frustration and persistence, but deep down, he knew progress was happening.

Months later, Adrian's teacher recorded one of his practice sessions and played it back to him. He was stunned by how much better he sounded compared to his first attempt. His brain had been quietly rewiring through repetition, strengthening connections between the auditory, motor, and sensory regions. Every mistake had been a signal to adjust, refine, and improve. The feedback from listening to himself accelerated the process, helping him recognize patterns and correct them.

As his confidence grew, so did his enjoyment. He found himself looking forward to practice, losing track of time while playing. The music began to flow more naturally, and what once felt mechanical became expressive. His teacher explained that this shift reflected a deeper level of neuroplasticity. The skill had moved from conscious effort to automatic execution. His brain had created dedicated pathways that allowed him to play without overthinking each movement.

One day, his daughter joined him for a duet. As they playe together, Adrian felt something profound. The years betwee them vanished, replaced by a shared rhythm and connectio The joy of learning had transformed into the joy of creatin He realized that his motivation—the desire to connect with h daughter—had been the fuel that powered his brain transformation. Every note was a reflection of the effor patience, and emotion invested in his journey.

Looking back, Adrian saw that learning the violin was nev just about music. It was about proving to himself that the adu brain is still capable of growth. His story became a remind that learning has no age limit and that skill development always possible when driven by purpose. He often told h friends, "Your brain is always ready to learn—you just have give it a reason." The violin became his symbol of persistenc the sound of neuroplasticity in motion.

Chapter 8: Neuroplasticity and Technology

Technology has become one of the most powerful tools for enhancing neuroplasticity. From brain imaging to virtual reality therapy, modern innovations allow scientists and clinicians to observe, stimulate, and even accelerate the process of brain rewiring. Where recovery and learning once relied solely on human effort, technology now provides feedback, precision, and motivation in ways that were once impossible. These breakthroughs have revolutionized rehabilitation, education, and the way we understand human potential.

Functional MRI (fMRI) and PET scans have given researchers unprecedented insight into how the brain changes. By measuring blood flow and activity, these tools show which regions of the brain engage during specific tasks. They have confirmed what was once theory: the brain physically reorganizes itself when learning new skills or recovering from injury. With this knowledge, therapy can now be tailored to each person's unique brain patterns, allowing for more personalized and effective interventions.

Virtual reality (VR) has emerged as a particularly promising tool in neurorehabilitation. VR environments immerse patients in simulated experiences that stimulate multiple senses simultaneously. For stroke survivors, this might mean practicing reaching or walking in a safe, controlled space. For those recovering from trauma, it allows exposure to positive experiences that retrain emotional responses. By combining visual, auditory, and tactile feedback, VR strengthens neural connections faster than traditional methods because it engages the brain's reward and motivation systems while promoting repetition.

Artificial intelligence is also transforming neuroplasticity-based therapies. AI-driven programs can analyze a patient's performance and adjust difficulty levels in real time, ensuring that exercises remain challenging but achievable. These systems create a personalized feedback loop that mirrors how the brain learns—by trial, error, and adaptation. AI is also being used to decode brain signals for communication devices, helping individuals with paralysis express thoughts and emotions through neural interfaces. This not only restores communication but also reactivates dormant neural networks, encouraging further brain reorganization.

Wearable technologies and mobile apps now make it possible for people to monitor and improve their cognitive and emotional health daily. EEG headbands, mindfulness trackers, and memory-training platforms measure brain activity and provide feedback that promotes consistent practice. Even simple tools, such as language-learning or puzzle apps, use repetition and reinforcement to encourage neuroplasticity. By transforming therapy into interactive, gamified experiences, these technologies make brain training accessible to millions of people outside clinical settings.

Technology's role in neuroplasticity is not to replace the brain's natural processes but to amplify them. The human brain remains the driver of change; technology simply provides the tools to guide and accelerate it. As our understanding of neuroscience deepens, the combination of human determination and technological innovation continues to unlock new levels of recovery and performance. From brain-computer interfaces to AI-driven learning systems, technology has made it clear that neuroplasticity is not limited by biology—it's enhanced by imagination.

Leah's story

When Leah suffered a stroke at fifty, she was determined to regain her independence. Traditional therapy helped, but progress was slow, and frustration set in. Her therapist suggested she try a new virtual reality program designed for stroke rehabilitation. Leah was skeptical at first, unsure how video games could help her walk again, but she decided to try. The first time she put on the VR headset, she found herself standing in a digital garden filled with light, sound, and colo A gentle voice guided her to reach for glowing orbs floating i the air.

At first, her movements were hesitant and clumsy. Her righ arm barely responded, and she felt awkward and embarrasse But the virtual garden responded instantly—each successfu reach triggered a burst of light and sound, rewarding he effort. The feedback made her smile, something she hadn' done during therapy in months. Over time, the session became more engaging. She began looking forward to her dail practice, eager to beat her previous scores. What once felt lik rehabilitation now felt like a game.

Weeks turned into months, and Leah's real-worl improvements began to match her virtual achievements. Sh could now lift her arm higher, hold objects longer, and eve manage simple tasks around the house. Her therapis explained that the VR system had been activating multip parts of her brain simultaneously—motor, sensory, an emotional areas—all working together to rebuild los connections. The technology had not performed magic; it ha simply harnessed her brain's own ability to adapt.

Encouraged by her success, Leah began using other digit tools to continue her progress. She downloaded a mindfulnes app that guided her through daily breathing exercises and a memory-training game that kept her mind sharp. Each piece of technology became part of her healing routine. She began to see her phone, tablet, and VR headset not as distractions but as allies in her recovery. Her home, once filled with frustration, now buzzed with the sounds of progress.

Months later, her therapist invited her to join a group session where patients used brain-computer interfaces to control digital objects with their thoughts. The first time Leah moved a small cursor across the screen using only concentration, she cried. The experience reminded her of how far she had come—from struggling to lift her hand to commanding a computer with her mind. The technology had given her a glimpse into the limitless potential of the human brain.

Looking back, Leah realized that technology had not replaced her effort; it had amplified it. The screens, sensors, and simulations were tools, but the real transformation came from within her. Her determination to heal, combined with technology's ability to guide her brain's plasticity, had changed her life. When she spoke at a rehabilitation conference years later, she said, "The machines didn't heal me. They helped me remember that my brain could." Her story became a testament to the partnership between human will and technological innovation in the journey of neuroplasticity.

Chapter 9: Barriers to Neuroplasticity

While neuroplasticity gives the brain extraordinary power t adapt, certain factors can slow, limit, or even reverse th process. Understanding these barriers is essential for creatin conditions that support healing and growth. The mo common obstacles to neuroplasticity include chronic stres lack of sleep, poor nutrition, emotional distress, inactivity, an negative thinking patterns. Each of these can disrupt th brain's chemistry and reduce its ability to form and strengthe new connections. Recognizing and managing these barriers the first step toward maintaining a brain that is capable c continual renewal.

Stress is one of the most significant inhibitors c neuroplasticity. When the brain perceives danger—wheth from real threats or constant worry—it releases cortisol an adrenaline. These hormones prepare the body for action bu can be damaging when sustained over time. Chronic stre floods the brain with cortisol, which impairs the hippocampu a region crucial for memory and learning. It also reduces leve of brain-derived neurotrophic factor (BDNF), a protein th supports the growth of new neurons. Without adequate BDN the brain's ability to adapt and rewire slows dramatically.

Sleep deprivation has a similarly harmful effect. During slee the brain consolidates memories, clears waste, and repairs neural connections. Lack of sleep interrupts these processes, weakening existing pathways and preventing the formation of new ones. People who consistently sleep fewer than six hours a night often experience decreased focus, emotional instability, and reduced learning ability. Sleep is not a passive state—it is one of the brain's most active periods of renewal, essential for maintaining neuroplastic health.

Nutrition also plays a crucial role. The brain relies on specific nutrients—omega-3 fatty acids, antioxidants, amino acids, and vitamins—to function optimally. Diets high in processed foods, sugars, and unhealthy fats can lead to inflammation that interferes with neural communication. Conversely, diets rich in fruits, vegetables, whole grains, and lean proteins provide the building blocks for neurotransmitters and myelin, which protect neural pathways. The saying "you are what you eat" applies profoundly to brain function. A nutrient-poor diet deprives the brain of the materials it needs to rewire effectively.

Inactivity and isolation create another set of barriers. Physical exercise increases oxygen and blood flow to the brain, stimulating the release of growth factors that encourage neural repair. Sedentary lifestyles reduce these benefits, while isolation deprives the brain of social and emotional stimulation. Human connection is a key driver of neuroplasticity because it activates regions associated with empathy, language, and problem-solving. Without regular physical and social engagement, the brain becomes less flexible, and cognitive decline can accelerate.

Finally, mindset plays a decisive role. Negative self-talk and fixed beliefs—such as "I can't change" or "I'm too old to learn"—can weaken motivation and prevent consistent effort, effectively blocking the rewiring process. The brain responds to expectation: when we believe improvement is possible, the body releases chemicals that support learning and growth. When we give up, those same systems shut down. Overcoming barriers to neuroplasticity therefore requires more than science; it requires faith in the brain's ability to change. The mind must lead the body toward healing.

Henry's will to change

After his stroke, Henry was determined to regain his independence. But weeks of therapy left him discouraged. His progress was slow, and his frustration grew each day. He barely slept, replaying his failures in his mind. His diet had deteriorated into fast food and coffee, and he avoided friends who wanted to visit. Though he didn't realize it, every one of these habits was quietly working against his brain's ability to heal.

One afternoon, his therapist noticed that Henry's motivation had faded. Instead of pushing him through another exercise, she sat down and asked about his life outside therapy. As he described his sleepless nights, constant anxiety, and loneliness, she explained that these factors were powerful barriers to neuroplasticity. "Your brain is trying to rewire," she said, "but stress and exhaustion are like noise in the system. We need to quiet the noise so your brain can focus on healing." Her words struck him deeply. For the first time, he realized that recovery wasn't just about physical movement—it was about creating the right conditions for his brain to grow.

With his therapist's help, Henry began to rebuild his daily routine. He started going to bed at a consistent time, replacing late-night worry with calming music and breathing exercises. His therapist encouraged him to walk short distances each morning, even if it was just to the end of the block. The movement brought him fresh air, sunlight, and a sense of accomplishment. He also began preparing healthier meals with the help of a neighbor who visited weekly. These small changes began to reshape his mindset.

As the weeks passed, Henry noticed subtle improvements. His concentration sharpened, and he began to feel more optimistic. He reconnected with old friends, and their visits filled his house with laughter. Each conversation became an exercise in both communication and emotional healing. Th stress that once clouded his mind started to lift, allowing hi therapy sessions to become more productive. He realized tha by reducing the barriers around him, he had given his brai the space it needed to rewire.

Months later, Henry's doctor showed him a new scan of hi brain. Activity had increased in regions that had been quie after the stroke. His recovery was no longer stagnant—it wa accelerating. He felt a wave of gratitude and pride. The ma who once believed he was stuck had become an example o what was possible when barriers were removed. His brain ha responded to his efforts with renewed strength.

Looking back, Henry saw that his greatest obstacle had no been the stroke itself but the stress, fear, and habits that ha followed it. By addressing those barriers, he unlocked hi brain's natural capacity to heal. He began speaking to othe patients in group therapy, telling them, "You can't grow a tre in a storm. Calm the wind first—then the roots will take hold. His story became a lesson in patience and the importance o nurturing the environment where neuroplasticity can flourish

Chapter 10: Building a Brain for Life

The science of neuroplasticity reveals one essential truth: the brain is a living system that continually reshapes itself in response to how we use it. Every thought, action, and emotion either strengthens or weakens neural connections. To build a brain for life—one that remains sharp, adaptable, and resilient—we must consciously cultivate habits that promote growth. This involves nurturing physical health, mental engagement, emotional balance, and social connection. A lifelong adaptable brain is not a gift of genetics; it is the outcome of intentional living.

Physical movement is one of the most powerful ways to sustain neuroplasticity. Exercise increases oxygen and blood flow to the brain, stimulates the release of growth factors like BDNF, and enhances the formation of new neurons. Activities that combine movement with coordination—such as dancing, swimming, yoga, or martial arts—are particularly effective because they engage multiple areas of the brain simultaneously. These activities challenge memory, balance, and focus while promoting physical health, creating a feedback loop that strengthens both body and mind.

Mental stimulation is equally vital. The brain thrives on novelty and challenge. Learning new skills, reading complex material, solving puzzles, or exploring creative outlets keeps the neural networks active and growing. Routine and predictability, while comfortable, can lead to mental stagnation. By exposing the brain to new experiences, we encourage the creation of new connections and prevent cognitive decline. A brain that continues to learn is a brain that continues to grow.

Emotional health plays a central role in maintaining neuroplasticity throughout life. Chronic negativity and stress can weaken the brain's adaptability, while positive emotions enhance it. Practices such as gratitude, mindfulness, and deep breathing help calm the nervous system and regulate the release of neurotransmitters. This emotional stability supports focus, motivation, and long-term learning. A balanced emotional state is like fertile soil—the brain's roots grow stronger when the environment is calm and supportive.

Social interaction also fuels brain growth. Conversations, teamwork, and empathy activate multiple neural systems simultaneously, including those involved in language, memory, and emotional regulation. Strong relationships provide emotional safety and mental stimulation, both of which reinforce healthy neural pathways. Isolation, by contrast, deprives the brain of the variety and feedback that needs to thrive. Staying connected—through family, friends, or community—helps the brain remain flexible and engaged.

The most important element of lifelong neuroplasticity is purpose. People who pursue goals that matter to them—whether personal, creative, or humanitarian—tend to stay mentally sharp and emotionally fulfilled. Purpose keeps the brain engaged by combining motivation, meaning, and persistence. It transforms effort into growth. Building a brain for life means choosing daily habits that feed curiosity, movement, connection, and joy. The goal is not perfection but progress—a steady commitment to nurturing the mind's endless potential.

Now, let's talk about David

When David retired at sixty-five, he imagined he would finally rest after decades of long hours and constant deadlines. But within a year, he began to feel restless and disconnected. His days blended together, filled with television and routine errands. He noticed that his memory was slipping—names, appointments, even familiar routes around town. One afternoon, while searching for his car keys for the third time that day, he realized that rest had become stagnation. He decided it was time to challenge his brain again.

He started small, enrolling in a photography class at a local community center. The first lessons were frustrating. The technical terms confused him, and the camera buttons felt foreign under his fingers. Yet the challenge awakened something in him. Each class required focus, patience, and creativity. As he practiced, he began to see improvement—not just in his photos but in his mental clarity. Remembering settings, framing shots, and editing images forced his brain to form new pathways. He felt sharper, more present, and more alive.

Encouraged by his progress, David added morning walks to his routine. He began exploring new trails, often taking his camera along to capture early sunlight filtering through the trees. The combination of physical activity and creative focus made him feel stronger and calmer. He learned that movement and mindfulness could work together, reinforcing both body and brain. His doctor noticed improvements in his mood, energy, and cognitive performance. David had not only changed his routine; he had changed his biology.

His social life grew as well. At the photography center, he made new friends who shared his enthusiasm for learning. They organized weekend trips to practice their skills, exchanging ideas and laughter. These connections brought warmth and meaning to his life. He found himself looking forward to conversations, discussions about art, and the thrill of seeing something beautiful through another person's eyes. The joy of community reactivated emotional and cognitive circuits that had been dormant for years.

Months later, David began volunteering at a local youth program, teaching photography to teenagers. Sharing his skills reignited his sense of purpose. Watching his students learn and grow reminded him of his own progress. His brain stayed alert as he adapted his teaching methods, remembered names, and guided each student's creative process. The sense of contribution and purpose brought him a fulfillment that retirement alone never could.

Looking back, David realized that he had built more than a hobby—he had built a brain for life. Through movement, learning, connection, and purpose, his mind had grown sharper and more resilient than it had been in years. He had learned that neuroplasticity doesn't fade with age; it responds to how we live. He often told his students, "Keep challenging your mind and sharing what you learn. That's how we stay young—inside and out." His story became a lasting reminder that the brain's ability to adapt is a lifelong invitation to grow.

Epilogue: The Living Brain

The human brain is often described as a universe containe within itself—a galaxy of billions of neurons, each pulsing wit the potential for change. Through the science c neuroplasticity, we now understand that this universe is nc static but alive, dynamic, and always becoming somethin new. Every experience, every thought, and every act c learning reshapes its landscape. The story of the brain is therefore, the story of evolution happening within each of us moment by moment.

The discoveries explored throughout this book prov something extraordinary: transformation is possible at an stage of life. The brain does not retire from growth. It listens responds, and adapts to how we live, think, and feel. This i not just a scientific truth but a human one. It means tha healing after injury, learning after failure, and renewal afte loss are built into our very biology. We are wired for resilience Neuroplasticity is not a rare phenomenon—it is our natura state.

To truly understand neuroplasticity is to see that every choic we make is an act of shaping the mind. When we practic patience, we strengthen the pathways of calm. When w nurture curiosity, we reinforce the circuits of discovery. Whe we forgive, connect, and love, we deepen the networks that sustain empathy and joy. The brain does not distinguish between physical and emotional practice; it records both as experience. Each thought becomes architecture, each feeling a brushstroke in the mural of who we are becoming.

But this knowledge also carries responsibility. Neuroplasticity means that we cannot remain passive observers of our own minds. The patterns we repeat—our habits, our reactions, our beliefs—determine the future shape of our brains. When we dwell on fear or self-doubt, those pathways grow stronger. When we cultivate optimism, gratitude, and purpose, they take root instead. Every moment offers a chance to redirect the flow of thought, to choose which circuits we strengthen and which we allow to fade. The sculptor of the brain is not time but intention.

The stories in this book have shown that even in the face of trauma or decline, the brain remains capable of astonishing renewal. A stroke survivor relearns to paint; a child with a learning disorder finds their voice; an elderly person discovers new joy through art. These are not miracles—they are the natural expressions of a system designed to adapt. The same processes that allow us to recover from loss also allow us to evolve into wiser, more compassionate versions of ourselves.

Neuroplasticity gives us the freedom to redefine what is possible.

The future of neuroscience continues to unfold, and with it comes a new vision for humanity. As technology merges with biology, as education evolves, and as medicine becomes more personalized, we are learning how to guide neuroplasticity deliberately. Yet the greatest discovery remains timeless: the most powerful tool for shaping the brain is still the human spirit. Curiosity, courage, and connection are the true drivers of transformation. When combined with knowledge, they form a bridge between science and the soul.

To build a better brain is to build a better life—and ultimately a better world. Our capacity to adapt is not limited to neurons; it extends to how we treat each other, how we respond to challenges, and how we imagine the future. The brain's flexibility mirrors the resilience of the human heart. Each time we choose growth over fear, compassion over indifference, and effort over surrender, we participate in the miracle of renewal that neuroscience has only begun to describe.

In the end, neuroplasticity teaches us that the brain is never finished—it is always writing, always rebuilding, always learning. Whether we are healing from injury, mastering new skills, or seeking meaning, our minds remain open systems capable of infinite transformation. The journey is ongoing, shaped by every action, emotion, and choice we make. The brain you wake up with tomorrow will not be the same as the one you have today. That truth, perhaps more than any other, is what makes being human so extraordinary.

www.ingramcontent.com/pod-product-compliance
Lightning Source LLC
LaVergne TN
LVHW022028200926
848481LV00007B/2590
* 9 7 9 8 3 4 9 6 6 2 6 3 8 *